Special Report:
Improving Firefighter Communications

Reporting by: Adam Thiel
Edited by: Hollis Stambaugh

This is Report 099 of the Major Fires Investigation Project conducted by Varley-Campbell and Associates, Inc./TriData Corporation under contract EMW-94-C-4423 to the United States Fire Administration, Federal Emergency Management Agency.

 **Homeland Security**

Department of Homeland Security
United States Fire Administration
National Fire Data Center

U.S. Fire Administration Fire Investigations Program

The U.S. Fire Administration develops reports on selected major fires throughout the country. The fires usually involve multiple deaths or a large loss of property. But the primary criterion for deciding to do a report is whether it will result in significant "lessons learned." In some cases these lessons bring to light new knowledge about fire--the effect of building construction or contents, human behavior in fire, etc. In other cases, the lessons are not new but are serious enough to highlight once again, with yet another fire tragedy report. In some cases, special reports are developed to discuss events, drills, or new technologies which are of interest to the fire service.

The reports are sent to fire magazines and are distributed at National and Regional fire meetings. The International Association of Fire Chiefs assists the USFA in disseminating the findings throughout the fire service. On a continuing basis the reports are available on request from the USFA; announcements of their availability are published widely in fire journals and newsletters.

This body of work provides detailed information on the nature of the fire problem for policymakers who must decide on allocations of resources between fire and other pressing problems, and within the fire service to improve codes and code enforcement, training, public fire education, building technology, and other related areas.

The Fire Administration, which has no regulatory authority, sends an experienced fire investigator into a community after a major incident only after having conferred with the local fire authorities to insure that the assistance and presence of the USFA would be supportive and would in no way interfere with any review of the incident they are themselves conducting. The intent is not to arrive during the event or even immediately after, but rather after the dust settles, so that a complete and objective review of all the important aspects of the incident can be made. Local authorities review the USFA's report while it is in draft. The USFA investigator or team is available to local authorities should they wish to request technical assistance for their own investigation.

For additional copies of this report write to the U.S. Fire Administration, 16825 South Seton Avenue, Emmitsburg, Maryland 21727. The report is available on the Administration's Web site at http://www.usfa.dhs.gov/

U.S. Fire Administration
Mission Statement

As an entity of the Department of Homeland Security, the mission of the USFA is to reduce life and economic losses due to fire and related emergencies, through leadership, advocacy, coordination, and support. We serve the Nation independently, in coordination with other Federal agencies, and in partnership with fire protection and emergency service communities. With a commitment to excellence, we provide public education, training, technology, and data initiatives.

ACKNOWLEDGMENTS

The U.S. Fire Administration greatly appreciates the cooperation received from the following people and organizations during the preparation of this report:

City of New York (NY)
Fire Department

Commissioner Thomas Von Essen
Assistant Chief Peter Ganci
Assistant Chief Frank Cruthers
Deputy Chief James Murtagh
Deputy Chief Stephen King
Deputy Chief Richard Fanning
Battalion Chief Edward Geraghty
Battalion Chief Richard Comiskey
Battalion Chief Frank Miale
Battalion Chief George Gierer
Lt. Steve Spahl – Special Operations Command
Members of Engine Companies 48 and 75
Members of Ladder Companies 37 and 33
Members of Rescue Company 1

Memphis Division
of Fire

Director Charles Smith

Other References

- TriData Corporation, *Wildland Firefighter Safety Awareness Study Phase III-Implementing Cultural Changes for Safety*, March 1998

- Varone, J. Curtis, *Fireground Radio Communication and Firefighter Safety*, National Fire Academy Executive Fire Officer Program, 1996 Outstanding Research Award

- Kaprow, Miriam Lee, *Magical Work: Firefighters in New York*, Human Organization (Society for Applied Anthropology, Vol.50, No.1, Spring 1991)

- Federal Aviation Administration, *Developing Advanced Crew Resource Management Training: A Training Manual*, August, 1998

- Comfort, Louise K., ed., *Managing Disaster: Strategies and Policy Perspectives*, (Durham: Duke University Press, 1988)

TABLE OF CONTENTS

Special Report:
Improving Firefighter Communications

EXECUTIVE SUMMARY

Several recent incidents involving firefighter fatalities demonstrate that, despite technological advances in two-way radio communications, important information is not always adequately communicated on the fireground or emergency incident scene. Inadequate communication has a definite negative impact on the safety of emergency personnel and may contribute to injuries or deaths of firefighters, rescue workers, and civilians.

Inadequate fireground communication is repeatedly cited as a contributing factor in many of the incidents reported through the United States Fire Administration Major Fires Investigation Project. This fact, coupled with the limited availability of research on such an important topic, prompted the United States Fire Administration (USFA) to study some of the potential causes of communication breakdown, and to provide recommendations that will help departments improve their operational communications.

While the findings contained in this special report are primarily oriented toward the municipal fire service, this does not diminish their potential relevance to other firefighters and emergency responders. With respect to communicating in high-stress environments, numerous parallels exist across public safety and related disciplines. Some of the ideas presented here are drawn from the experiences of wildland firefighters, airline flight crews, and military personnel.

SUMMARY OF KEY ISSUES

Issues	Comments
Unsuitable equipment	The chief communication problem reported by firefighters and company officers is the difficulty with communicating from inside a fire when using full personal protective equipment, including SCBA. The majority of portable radios currently used by fire departments are ill-suited for the task.
Portable radios needed for all firefighters	Despite some technical limitations, portable radios are a proven lifesaver during fires and emergency incidents, and should be considered a critical item of personal protective equipment akin to SCA. Ideally, **every firefighter working in a hostile environment should have a portable radio with emergency distress feature.**
Little attention paid to human factors	There is a dearth of available literature pertaining to the impact of human facts on effective fireground communication. Furthermore, while fire departments devote substantial time to manipulative skill training, relatively little training is provided to help firefighters develop stress-tempered communication skills.
Importance of active listening	All firefighters on an emergency incident should actively monitor their radios for important information at all times, not just when specifically queried. Communications should be emphasized as an essential part of firefighter function as a tactical team, not just operating as individual entities.
Standard message formats and language	Fire departments can enhance fireground communication by creating standard message formats and keywords used consistently. Plain English is usually preferred over codes, especially when transmitting a complex message.
Tiered message priority	Keywords to prompt immediate action can be tiered based on their priority, for example, "Mayday" signals a life-or-death situation, while "Urgent" may be used to signify a potentially serious problem. Such message headers prompt the crew's listening priorities and radio discipline.
Attention to cultural factors	If necessary, firefighters are not usually reluctant to circumvent the chain of command to report critical safety issues. There may be greater hesitation to communicate problems in completing an assigned task. However, this is usually due to a lack of situational awareness, and not a fear of reprisal from other members. Studies on firefighter communications show that sometimes the culture of bravery in the fire service is reflected in a reluctance to communicate quickly enough when help is needed. Repeated situations where this occurs should be closely examined by the fire officer.

INTRODUCTION

Communication problems are continually cited as contributing factors in fires and emergency incidents where firefighters are killed or injured. The number of "near-miss" incidents where fireground communication was ineffective may be higher than generally realized. The purpose of this special report is to help fire departments improve their communication processes to enhance scene safety, help prevent firefighter deaths and injuries, as well as to promote effective tactical operations.

Despite the obvious importance of effective communication on the emergency scene, only a limited amount of published research exists dealing specifically with this topic. Several recent programs have focused on improving wildland firefighter safety. A rich source of literature on communicating in emergencies is available in the field of airline crew resource management. Efforts to improve operational communication and coordination also exist in the military. This report will draw parallels between communications-related issues in these areas and the municipal fire service.

To validate the applicability of the information presented in this report on fireground communication, field research was conducted with assistance from the City of New York Fire Department (FDNY). The FDNY was selected based on frequency of fire incidents, a strong interest in the topic, and a demonstrated willingness to cooperate with USFA researchers. While certainly not the "final word" on fire service communication, this special report may help identify future avenues of inquiry and will provide some suggestions that can be implemented by fire departments.

INCIDENT SUMMARIES

Many fire departments conduct formal internal reviews of their major incidents to enhance their training for better practices. However, the U. S. Fire Administration examines major incidents to ascertain "lessons learned" that can be communicated to the nation's fire safety community. Summaries of communications-related lessons learned at a variety of incidents are summarized below to help identify commonalties and trends in the experience of fire departments.

1. *Wood Truss Roof Collapse Claims Two Firefighters, Memphis, TN (USFA Technical Report Series)* Two firefighters operating an interior attack line were killed at this fire after a church roof collapsed. "Communications was also a problem, contributed to the lack of organization at the scene, since the Incident Command was unable to communicate with company officers on the tactical radio channel." (p. 21)

2. *Indianapolis Athletic Club Fire, Indianapolis, IN (USFA Technical Report Series)* Two firefighters were killed on the third floor of a nine-story, mixed use building during a fire. "New radio equipment and lack of familiarity with its operation may have contributed to delays in acknowledging and processing requests for additional companies." (p.28) "Radio discipline is important." (Appendix E) "His most serious injuries were a direct result of having to compromise his personal safety in order to send a distress signal. While the push-to-talk switches could be operated with minor difficulty with a gloved hand, the emergency or distress button is virtually impossible to operate in the same manner." (p.27)

3. *Four Firefighters Die in Seattle Warehouse Fire, Seattle, WA (USFA Technical Report Series)* Four firefighters were killed in a collapse at an arson fire in a warehouse. "The interior attack crews on the upper level did not report that very little fire had been found inside the building and that all flames appeared to have been knocked down. The crews on the lower level did not report that they had found a large area that was fully involved in fire. The discrepancy between these reports would have alerted the Incident commander to reevaluate the attack plan." (p.4)

4. *Two Firefighters Die in Auto Parts Store Fire, Chesapeake, VA (USFA Technical Report Series)* Two firefighters were killed after a roof collapse in a retail auto parts store. Poor communications was one of the problems that investigators determined contributed to the firefighters' death. According to the report, "the fireground operations were conducted on the same radio channel as the routine dispatch and transfer of additional units, hampering the fireground communications during the early stages of the incident." The Chesapeake Fire Department recently upgraded their communications equipment and "added additional portable radios to each piece of apparatus' to shore up their communications interoperability.

TYPES OF COMMUNICATION PROBLEMS

Communication problems commonly encountered by firefighters are broadly divided into two categories. First are those problems related primarily to mechanical/technical issues such as unsuitable equipment, radio malfunction, limited system capacity, or atmospheric interference. The second category of problems is somewhat broader and includes the critical human factors necessary for effective communications, for example, radio discipline and completing the communications "loop". While the research literature dealing with fire service communication overall is sparse, more has been written about the technical aspects of the issue than about the human factors. This special report addresses both, with particular emphasis on the critical human factors involved in improving firefighter communications.

Technical Issues

There are a variety of technical issues that may affect communications among firefighters and emergency personnel. While the applicability of these specific challenges to individual departments may vary depending on multiple factors, the advance of technology holds promise for solving some of the technical problems commonly encountered today.

Unsuitable Equipment

The predominant communications-related concern reported by firefighters and company officers is the difficulty in communicating while using self-contained breathing apparatus (SCBA). The need for SCBA in hazardous environments is well understood by firefighters. By the same token, firefighters engaged in fighting a fire are acutely aware of the importance of effective communication for tactical and strategic decisionmaking, coordination among different units, and for transmitting urgent safety-related messages.

The use of SCBA, while critical firefighter safety, can interfere with effective communication, both face-to-face, and via portable radio. Few firefighters are unfamiliar with this problem and most have asked, "What did they just say?", after attempting to comprehend a radio transmission sent by an interior firefighting crew. Even face-to-face conversation through SCBA is extremely difficult during a working fire due to high levels of background noise and the barrier imposed by the facepieces.

Fireground safety concerns dictate that firefighters *both* use SCBA *and* communicate effectively. This can create a dilemma when vital messages must be clearly communicated within the fire environment. As a result, some firefighters report they have found it necessary to momentarily remove their SCBA facepiece in order to transmit a message over a portable radio or directly to a colleague. The obvious danger of removing the SCBA facepiece, even for a brief moment, is evident considering the thermal, toxic, and oxygen-deficiency hazards posed by a fire and the resulting products of combustion. A single unprotected breath of such an atmosphere may be sufficient to cause long-term health problems, incapacitation, or even death. Firefighters are certainly aware of this risk. Since they repeatedly expose themselves to the potential for serious injury in order to effectively communicate, there is a clear need to continue improving technology to correct current systems limitations.

SCBA manufacturers are cognizant of the problems faced by firefighters attempting to communicate through properly used SCBA. A variety of products are currently available that seek to mitigate the problem including speech ports, facepiece-integrated microphones, intercom systems, portable radio interfaces, throat mikes and "bone" mikes worn in the ear or on the forehead. Most of these systems have received relatively mixed reviews from firefighters in the field and the current cost of the most effective, systems may be prohibitive for many fire department budgets. The absence of effective affordable equipment suggests that more work needs to be done to develop a durable, easy-to-use system for enhancing voice communications in conjunction with properly worn SCA.

The unsuitability of currently available portable radio equipment for use in hostile fire environments was another recurring theme in this research. Firefighters are almost universally dissatisfied with their portable radios (irrespective of make or model), under those conditions. The problems are ones chiefly related to ergonomics and durability. The trend toward miniaturization allows radios to contain more features in smaller, more lightweight packages. From an ergonomic standpoint however, the switches and dials on the newest radios tend to be to small to operate with a gloved hand. Additionally, many of these radios have Liquid Crystal Display (LCD) screens that, although backlit, can be difficult to read in a low-visibility environment. An example of ergonomic limitations was seen during a fire in Indianapolis in which two firefighters were killed, and a captain was severely burned after he removed his glove to operate an emergency signaling device on his portable radio. The captain had repeatedly attempted to activate the ESD feature on his radio, but could not do while wearing full personal protection equipment.

Ideally, radios used for firefighting should be highly water-resistant, shock-resistant, and designed for easy operation by firefighters wearing heavy gloves in a hostile environment. Currently, such a radio does not exist. The process of adapting other technology for fire service use has brought as the units and equipment used today. The needs articulate by firefighters suggest a unique enough requirement to engineer an effort "designed around" their specific environment.

The potential for linking video, audio, and other data feeds from firefighters operating on the interior of a structure to outside monitors, also holds promise for enhancing fireground communication. Technology transfer efforts involving various universities, government agencies, and fire departments are underway to develop model systems. Realistically, the deciding factor in the ultimate implementation of these technological adjuncts, as well as for improving communication through SCBA and developing a better portable radio for firefighting, is cost.

Despite fire service limitations with the current types of portable radios, it is highly recommended that *every* firefighter entering a fire situation be equipped with a portable radio. Ideally, the radio should be worn where the channel selectors, display, and emergency signaling device-a vital safety feature-are readily accessible to the firefighters without breaching the personal protective envelope (e.g., in a radio pocket or water/shock-resistant case outside the coat). As with any tool, firefighters must receive training in the proper operation, limitations, and preventive maintenance of the portable radio. It is also recommended that battery chargers and spare batteries be kept onboard fire apparatus to ensure radios are fully charged at all times. The members of FDNY Rescue Company One have such a setup and regularly change the batteries in their portable radios between calls, and even during runs where they expect to use their radio for an extended period of time, such as on a "working fire". Battery life and recovery periods are also important considerations for departments that use portable radios on a relatively infrequent basis since even unused batteries weaken over time.

There are some things individual firefighters can do to enhance communications through SCBA that do not require technological adjuncts. The voice ports currently integrated into several manufacturers' SCBA facepieces work best when firefighters speak in a normal to moderately loud tone of voice. High-pitched sounds do not transmit well through these devices. Therefore, firefighters attempting to communicate through SCBA facepieces so equipped, should attempt to speak calmly, at moderate volume, and with clear enunciation. When using these facepieces to transmit via portable radio, it may help to hold the remote microphone directly in front of the voice port. Face-to-face conversation through SCBA may be aided by placing the facepieces of sender and receiver close together which has the added benefit of facilitating physical and eye contact between the parties. Clearly, in an environment where visibility is limited, this is not a viable option.

Although not formally recommended, some firefighters report better success during portable radio communications, while using SCBA if they transmit with the remote microphone placed directly again their neck, instead of holding it in front of the facepiece. The effectiveness of this technique may vary depending on the characteristics of the microphone, the user's anatomy, and the scene conditions. Throat microphones are specifically designed for sound transfer based on concept, but are not common in fire service applications.

Equipment Failure

Modern public safety radio communication systems are complex and highly technical. They may encompass a multitude of fixed antenna sites, Computer Aided Dispatch (CAD) terminals, mobile radios, Mobile Display Terminals (MDTs), portable radios, cellular telephones/faxes/modems, and even laptop computers. The proper functioning of this equipment is of paramount importance to firefighters and rescue personnel operating on the emergency scene. For example, many of the 800mHz radio systems currently in use depend upon technology to the extent that, should one part of the system fail, even line-of-sight radio communications can be negatively impacted. Obviously, firefighters operating on the fireground should not have to worry about the functionality of the overall communication system.

To prevent equipment malfunctions from hampering effective radio communication, systems should be designed, installed, and maintained only by qualified technicians. Regular preventive maintenance will help minimize the occurrence of failures and routine radio checks are recommended for fire departments that do not use their radio equipment on a daily basis. Many departments conduct regular test of all radio equipment system components, and alternative power supply capabilities to assure their serviceability.

Inadequate System Capacity

There have been several instances where inadequate capacity of the radio system was deemed a contributing factor in incidents with negative outcomes. This is most likely to occur during complex, multi-alarm incidents with many units operating, and attempting to communicate, simultaneously. The sheer volume of radio traffic in this case may overwhelm dispatchers and the Incident Commander, and may prompt firefighters to turn down their portable radios to remove the continual distraction. There have been numerous instances where vital transmissions went unheard due to the volume of radio traffic on a system where capacity was exceeded. Although radio discipline, which will be discussed in detail later, may help minimize this problem, radio communication systems with multiple channel capability are best suited for fire department and emergency operations. This need was graphically illustrated by the experience of the Hackensack, NJ Fire Department at a 1988 fire that killed five firefighters. At this fire, numerous calls for help were transmitted by trapped firefighters and went unheard (or were overridden) due to excessive traffic on the single frequency radio system.

A dedicated dispatch channel is most often used to conduct routine communications operations. To prevent routine radio traffic from interfering with incident-specific communications, active incidents may be assigned to other channels for tactical operations, according to criteria established by the agencies involved and determined by the available capacity. Modern, "trunked" radio systems may have enough available frequencies for each incident to be assigned a separate tactical channel. Multiple-alarm fires or complex incidents like those involving hazardous materials or technical rescues may require multi-channel operations. Some departments, like the FDNY, regularly implement a command channel, separate from the fireground tactical channel, solely for the use of command-level officers at major incidents.

While the use of multiple channels for emergency operations is desirable, there are several important precautions that will help prevent problems from arising out of their use. Training is of vital importance to help familiarize personnel with using multiple channels on an incident and to identify potential problems. Unfamiliarity with the use of new radio equipment in Indianapolis was cited as a contributing factor in the casualties at the Indianapolis Athletic Club fire. Frequent utilization of the more complex, multi-channel systems during drills and routine operations will help enhance effective communication during unusual events.

Incidents where firefighters perished while calling for help on unmonitored channels indicate the need for continual monitoring of all channels in use during an operation. At the Regis Tower highrise fire that claimed the lives of two Memphis firefighters, several transmissions were made on alternate channels. The Incident Commander was unable to monitor both the dispatch and fireground channels simultaneously, while at the same time attempting to manage a difficult fire scene, and therefore missed the urgent calls for help from a trapped firefighter.

There are several possible ways to ensure that effective monitoring occurs. The FDNY dispatches a Field Communications Unit to multiple-alarm incidents, bringing dispatchers "to the street" and effectively shortening the length of the communications chain. Many other fire departments and emergency agencies maintain such units for use at major incidents. It may be helpful to utilize them more frequently. The FDNY also dispatches an additional battalion chief on all working fires to act as a communications coordinator. Fire departments choosing to follow this example can develop SOPs describing the duties and responsibilities of the communications coordinator at the incident scene.

Reference Standards

Several National Fire Protection Association (NFPA) standards address fire service radio communications including the 1992 Edition of NFPA 1500, *Standard on Fire Department Occupational Health and Safety Programs*; the 1990 Edition of NFPA 1561, *Standard for Fire Department Incident Management System*; the 1994 Edition of NFPA 1201, *Standard for Developing Fire Protection Services for the Public*; and the 1994 Edition of NFPA 1221, *Standard for the Maintenance and Use of Public Fire Service Communication Systems*. These standards address a variety of issues including the establishment of SOPs for communications personnel, the number of channels required for fire department radio systems, staffing of communications centers, and other related issues. Personnel responsible for the supervision and operation of fire department radio communication systems can use these NFPA standards to help ensure the adequacy of their systems. No NFPA standard currently addresses the certification of portable radio equipment intended for use during interior firefighting operations.

Interference

Atmospheric, environmental, and electronic interference may hamper effective communication at the incident scene. This type of interference can take many forms, ranging from the "skip" created by solar disturbances and atmospheric fluctuations; to interference caused by topographic factors like hills or tunnels. Communicating in high-rise buildings or ships is often difficult. While responding to an incident, radio transmissions may be compromised by the background noise of sirens. Obviously, some of these interfering factors cannot be controlled. Others, however, can be mitigated with some foresight.

Atmospheric interference caused by cosmic events is a fact of life and cannot be readily overcome by firefighters or other responders. However, personnel responsible for overseeing the operation of public safety communications systems should be alert to significant events that may cause interference of greater magnitude or duration than commonly encountered. This phenomenon results from solar disturbances on the surface of the sun. These disturbances known as "Solar Flares" are sudden releases in the solar atmosphere, which emit large volumes of electromagnetic radiation and highly energized atomic particles. Traveling at nearly the speed of light these highly energized particles pass through the ionosphere where they can affect satellite and radio communications on earth. These ionospheric irregularities can have adverse effects on radio signals over the entire frequency spectrum, however, generally having a great affect on radio frequencies above the 1GHz band range. Depending on the size and intensity of the Solar Flare, the ionospheric effects may linger a day or longer before subsiding.

Fire Department communication personnel should be conscious of this problem and be prepared to promptly address the issue when it arises. Although relatively infrequent, such atmospheric conditions could pose a communication problem during emergency operations. The Fairfax County (VA) Public Safety Communication Center recently sent a message to all systems users warning of the possibility for higher than normal levels of interference resulting from a solar flare as had been predicted by the National Oceanic Atmospheric Administration (NOAA). Fairfax County Fire Department maintains an Emergency Services Group which is responsible for monitoring NOAA at www.sec.noaa.gov for conditions relating to weather and atmospheric anomalies.

Interferences caused by atmospheric irregularities can be mistaken as an equipment problem or failure and can result in vital communication devices being placed out of service for repair. This misinterpretation of the problem could have far reaching implications for emergency personnel during emergency operations by reducing the usable number of serviceable radios available to personnel. The Federal Aviation Administration, in an effort to avoid such problems notifies all commercial air carriers of problematic atmospheric conditions.

Interference caused by topographic factors or characteristics of the built environment can often be identified and possibly corrected before related problems occur. Some model building codes recognize the fact that radio communications are difficult in high-rise buildings and therefore require the installation of hard-wired telephone systems for use by emergency personnel. In some urban areas, advanced radio technology and the installation of multiple antenna sites have alleviated the interference problems posed by large buildings. Fire departments can often identify problems in large area buildings during plans review or construction, and can work with developers to help ensure the adequacy of emergency communication systems in these structures.

Background noise is a common factor interfering with effective communication. Road noise and the sounds created by audible warning devices may drown out important radio traffic or make transmissions difficult to understand. The fireground or emergency scene can also be quite noisy due to the operation of fire pumps, hydraulic rescue tools, hoselines, and the noise created by people working in a stressful environment. Fortunately, there are some simple steps firefighters can take to alleviate the problems caused by these types of interference.

Depending on traffic conditions and applicable laws, it may be possible to use audible warning devices intermittently during a response. Sirens can be temporarily silenced before transmitting critical information over the radio. Electromechanical sirens should be equipped with a "brake" allowing them to be silenced relatively quickly. Closing the windows on some fire apparatus can help prevent road or siren noise from interfering with radio traffic. Noise attenuating devices like headsets further reduce ambient noise and can be equipped with intercoms and/or two-way radio interfaces. These devices serve several useful purposes. First, they limit outside interference, leading to clearer radio transmission and better comprehension of received messages. They enhance communications among crew members during a response and allow members to receive orders and review assignments prior to arrival on the scene, therefore reducing "reflex" time. Finally, they enhance overall safety by protecting crew members from hearing damage caused by exposure to high levels of noise (e.g., siren noise, engine noise, road noise).

While operating on the emergency scene itself, firefighters can take several steps to minimize interference. "Feedback", the squealing noise emanating from portable radios when located close together, can be reduced with awareness. When firefighters notice a colleague preparing to transmit a message, they can use their body to shield their own microphone, thus reducing the potential for feedback. This is as simple as turning to the side or covering the remote microphone with your hand while another member is transmitting. Keeping the volume on all radios turned to a moderately-loud level with also help prevent feedback. White it is important that the volume be kept high enough to hear all radio feedback. White it is important that the volume be kept high enough to hear all radio traffic, common sense dictates that maximum volume is inappropriate for all situations and may actually be distracting.

Firefighters can also help improve radio communications by maintaining awareness of the surrounding environment and its impact on radio equipment. The microphones on modern portable radios are very sensitive. If possible, command posts should be located away from noise sources (e.g., fire pumps, diesel engines, rescue tools). Some fire departments equip key personnel, including command officers and pump operators, with noise attenuating headsets to help ensure they hear vital messages. Recent experiences where firefighters' calls for help could not be heard lend credence to this practice. Members equipped with portable radios may be able to move away from high-noise areas when transmitting or receiving critical messages. One potential area of difficulty can occur when attempting to communicate near an activated PASS device. Upon locating a downed firefighter with PASS device sounding, the rescue team may need to deactivate the device to effectively communicate using their portable radios.

Human Factors

Although the technical aspects of fire service radio communications receive a good deal of attention in the literature, less attention is paid to the human factors affecting communication among firefighters and rescue personnel on the incident scene. In some cases the distinction between technical and human factors is difficult to make, however, it seems clear that human factors are critically important for ensuring safe and effective fireground communications. Good human communication skills and procedures will help promote safety even in the face of technical difficulties

Radio Discipline

Radio discipline is vital for effective communication among firefighters, dispatchers, and other emergency personnel. As mentioned previously in this report, a lack of radio discipline can overwhelm even robust communication systems, which still have finite capacities. Systems with inadequate capacities can become quickly overwhelmed even during routine incidents, seriously compromising firefighter safety. Allowing unlimited message transmission may create a situation where vital messages cannot be heard due to the number of less important transmissions being broadcast. By contrast, restrict radio traffic to only "vital" messages may prevent important information from being broadcast. The challenge, therefore is achieving a balance to ensure that all potentially important information is broadcast, but not at the expense of emergency transmissions or "Mayday" calls from interior crews.

It has been suggested by some in the fire service that the recommended practice of equipping all members with portable radios may exacerbate the radio discipline problem, possibly negating any positive safety effects. The experience of the City of New York Fire Department seems to allay these fears. Although all members are not yet equipped with portable radios, the FDNY currently deploys a much larger force of radio-equipped members on the fireground than in the past. When questioned, members of the FDNY at every organizational level emphatically denied that the larger number of portable radios on the emergency scene has created a severe radio discipline problem. On the contrary, all of those interviewed believe that the greater number of portable radios on the fireground has vastly improved firefighter safety and the quality of tactical decision-making.

There are several things firefighters can do to help improve radio discipline. An obvious way is to not use radios for communicating when face-to-face dialogue is a better and available choice. For example, when the sender and receiver are located a short distance from one another, when conferring about strategic or tactical options, or when a complex, vital message-such as a change in strategy from offensive to defensive-must be conveyed. Face-to-face communication is generally more effective than radio communication anyway, since both sender and receiver have the added benefit of using non-verbal cues to help convey ideas or understanding (e.g., eye contact, physical contact, body language). Distractions are also reduced and people can ask questions or identify problems more readily during one-on-one dialogue. Command officers can use runners to deliver and obtain information from remote units. Using a runner has the potential added benefit of providing another view of the situation to the Incident Commander.

Some incident management systems strongly recommend that transfers of command occur face-to-face, when possible. Several command-level officers in the FDNY expressed their preference for using face-to-face communication and runners wherever possible on the incident scene. Nevertheless, radio communication remains the most prevalent form of fireground communication currently in use.

Radio communication skills are critical for effectively conveying information at the incident scene. One of the most critical of these skills is being a good listener. Although it is often difficult to listen to radio traffic while performing fireground tasks, it is an important skill to develop. By doing so, firefighters can avoid re-broadcasting non-urgent messages that have already been transmitted and maintain awareness of the overall situation. Listening skills also help firefighters recognize when potentially urgent information has *not* been broadcast, and ensure that urgent messages are effectively communicated to the Incident Commander.

Good speaking skills are also vitally important for effective communication. Messages need to be transmitted using a logical format, at the appropriate volume, with good enunciation, and at a moderate pace. Most firefighters are familiar with the frustration of trying to understand someone either screaming or whispering into the radio, or an individual who speaks very fast, or too slowly. Prior to transmitting a message, firefighters should collect their thoughts and format the message in their head. Messages should be clearly stated without distracters like, "um", or "uh". Messages that are clear, direct, and to the point minimize unnecessary radio traffic and help prevent urgent messages from being delayed or unintentionally overridden.

The best way to develop good listening and speaking sills is through training and continued practice during multi-company operations drills or simulations. It may also be helpful for command or training officers to use tapes of actually incidents, or drills, to critique procedures and reinforce the importance of these skills. This can be done privately and will allow radio users to hear themselves, providing vital feedback for improvement. A FDNY battalion chief in the Bronx hold regular in-station radio communication drills using incident scenarios and radio-equipped members spread throughout the building. The resulting conversations are taped and played back during a critique to emphasize key learning points.

Another significant way to improve radio discipline is for the fire department to create SOP's describing standard message formats and distinguishing routine messages, urgent messages, and Mayday messages. In addition, standard terms should be defined for use during radio communication to help eliminate potential confusion and promote brevity during message transmission.

Situation Reporting

Accurate, regular situation reporting is critical for sound decision-making and for ensuring fireground safety. Command officers need regular situation reports so that they can make sound strategic and tactical decisions. These reports are often given by portable radio equipped members who, besides their obvious tactical duties like searching for victims and advancing hoselines, serve as the forward eyes and ears of the Incident Commander (IC). Without such information the IC may have to make decisions based on limited, incomplete, or inaccurate data.

Ultimately, it is the collective responsibility of every individual with access to a radio to ensure that radio discipline is maintained during emergency operations.

Incident Management

Most fire departments and public safety agencies regularly utilize some type of system for managing incident operations and personnel accountability at fire or emergency scenes. Many of the issues identified elsewhere in this report directly impact the proper functioning of a fire department's incident management system. Effective communications can help minimize potentially negative consequences to incident management and accountability brought on by rapidly changing situations, several of which are discussed here.

Fire department incident management systems and operational SOPs often detail assignments for specific units based on their order of arrival at the scene. For example, the first-in engine secures a water supply and begins fire attack, the second-in engine pumps the first engine's supply line, the first-in truck company conducts a primary search, etc. Proper coordination of these functions is vital to protect firefighters and effectively fight the fire. Achieving the required coordination is a function of effective communication and appropriate command and control practices. At a fatal fire previously investigated by the USFA, several companies on the initial alarm arrived "out of their assigned order" due to ice conditions on the hilly streets of the neighborhood. As a result, there was some confusion about what units were performing which tasks. Units arriving "out of order" and/or performing unassigned tasks should communicate this information to the Incident Commander so that adjustments can be made.

Coordination among units can be more difficult when mutual-aid or inter-agency responses are required. Incompatible communication nets are a common technical and tactical problem. As a result, units responding into surrounding jurisdictions may be equipped with different radios or a "patch" may be used to link different frequencies for the duration of an incident. With such efforts these may limit the technical problems associated with inter-agency responses, it may still be difficult for mutual-aid units to integrate themselves into an existing incident management system without effective communication. Incident Commanders may use terminology or expressions that are unfamiliar to personnel from other agencies. When receiving a message with unfamiliar terms, firefighters should clarify the message prior to taking action. Some fire departments dispatch a chief officer on all mutual-aid responses to serve at the command post and identify communications challenges before they cause operational problems.

Interpersonal communication is just as important as interagency communication for proper functioning of the incident management system. While the positions detailed in most systems are not specific to certain individuals, chief officers, company officers, and firefighters become use to working with the same people over time. When a newly promoted, "detailed", or "acting" officer or firefighter is on the scene, it can temporarily upset the previously existing working (and communication) relationships among personnel. Effective communication is the key for preventing negative outcomes. This communication must begin before an incident occurs. Department members at every organizational level should clarify expectations, reporting requirements, and any specific operations "nuances" as soon as possible upon recognizing that the new officers or firefighters are on board. This "feeling out" process is often performed subtly as members poll new people by asking about past experiences or relating their own stories.

Incident Commanders at complicated fire or emergency incidents will need support to ensure the effectiveness of fireground communications. It is not possible for a single individual to manage the scene, ensure accountability, make strategic and tactical decisions, and monitor one or more radio channels. Even at routine fires the potential for information overload is very real. Chief officers should be provided with aides early in the incident to help them manage communications and other tasks. Multiple aides may be needed to monitor radio traffic if several radio channels are used simultaneously. Some departments regularly assign aides to chief officers, while others allow for firefighters to be assigned, when needed, at the scene. Dispatching additional chief officers to working incidents can help alleviate the communications burden on the Incident Commander.

The number of significant incidents where urgent messages went unacknowledged points to the responsibility of everyone on the scene to actively listen to the radio communications for key words like "Mayday", "Urgent", and "Priority". If unclear whether or not these messages were correctly heard by the Incident Commander, firefighters should not hesitate to report the message to a sector officer, or directly to the command post is an especially important duty for members of Rapid Intervention Teams (RIT) and Firefighter Assistance (FAST) Teams.

Cultural

A great deal of research has been conducted recently to describe the culture of the wildland fire service and the impact of cultural factors on operational and safety issues. While it would undoubtedly be helpful to conduct an in-depth study of municipal fire service culture, such an endeavor is beyond the scope of this special report. Still, inasmuch as cultural factors influence effective fireground communication, they cannot be ignored.

Fire departments, and even individual fire companies, often have unique cultures arising from the particular nature of their work environment. While these cultures may vary somewhat in their specifics, a general set of values does exist within the municipal fire service as a whole. These values sometimes conflict with the mindset necessary to willingly communicate problem situations. The fire service culture typically emphasizes aggressiveness, an action, and the ability to overcome obstacles in the course of mission performance, while at the same time upholding a hierarchical organizational structure for fireground decisionmaking.

Chain of Command

Traditionally, fire department communications have been predominantly one-way. Emphasis is placed on "giving orders", "following orders", and "sending" messages. This is perhaps related to the traditional emphasis on unity of command and span of control as the primary means of maintaining order on the fireground. Although there is little room for extensive conversation on the emergency scene, the emphasis on maintaining the chain of command has created a potential communications problem. Firefighters may be reluctant to circumvent the chain of command and risk being considered insubordinate.

Interviews conducted with members of the FDNY seem to indicate that firefighters have minimal reluctance to communicate directly with chief officers when obvious safety issues are involved. A more common problem expressed by some command officers, is that firefighters report information to the wrong person because they are unaware of changes that were made to tactical assignments. When such misrouting occurs, it is important that the message recipient first relay the message to the appropriate person, and then advise the sender of the proper reporting pattern.

Problem Reporting

In some departments the culture of "heroism" attaches a stigma attached to calling for reinforcements. Where a flawed concept of bravery exists, firefighters often delay requests for help as long as possible to avoid being stigmatized. The potentially negative effects of this practice are obvious, as firefighters are trained to stay ahead of the fire at all times. Size-up is a continual process and requests for assistance should be transmitted as soon as a situation indicates help may be needed. Delay risks the possibility that reinforcements will arrive too late to influence the outcome of the situation. This has potentially negative consequences for both citizens, whose lives and property may be lost, and firefighters, who may be forced to work for an extended duration without relief. Members of the FDNY for example, understand well the physical stresses involved with firefighting and the potential for situations to change rapidly. Therefore, company and command-level officers are encouraged to call for additional resources as soon as possible. Indeed, additional resources are automatically dispatched upon the report of a "working fire" in many occupancies to help ensure adequate resources.

Depending on the capacity of the communications system and the adequacy of available radio channels, some Public Safety Agencies may designate specific radio frequencies as "Talk Around Channels." Such channels have no repeating capability thus limiting their transmitting range usually to a quarter mile or less. This allows emergency personnel to communicate freely with each other without fear of overloading the entire communication system. Incident Commanders can monitor the designated "Talk Around Channel" while remaining on a general fire ground channel. The primary channel frequency has message repeating capability. This kind of radio configuration allows the Incident Commanders the flexibility to simultaneously monitor both the fire ground activities and the communications dispatcher without switching channels. Also, as a result of the limited range of the "Talk Around" capability emergency personnel must direct all communications and request through the Incident Commander.

A related issue is the tendency for firefighters not to report problems completing an assignment; for example, forcing entry, procuring a water supply, or searching the fire floor. Firefighters sometimes are reluctant to report difficulties for fear of being judged as slow, incompetent, or unaggressive, all of which are contradictory to the fire department cultural values. While this may occur on occasion, the prevailing view from research conducted for this report is that firefighters operating inside a fire environment often lack the situational awareness to understand exactly how much time is passing, or to be cognizant of activities around them. Given the extreme nature of the fire environment, it is easy to understand how firefighters become prone to "tunnel vision". Effective communication is especially important here since the coordination required to safely accomplish fireground tasks may be compromised. When, after an appropriate period of time, the Incident Commander does not receive a situation report or keyword indicating completion of a critical task, for example an "all clear" on the primary search, the commander should query the assigned company for a report. Other company officers should listen to radio traffic while performing their assignments to ensure that activities around them will not have adverse impacts. For example, the officer of an engine company assigned to attack a fire should actively listen for any indication that the companies assigned to ventilation duties are having difficulty.

A tiered system of keywords should be used to prioritize the urgency of critical messages. For example, "Priority" messages may be defined as those requiring a swift response without the implication of immediate danger. "Urgent" may denote a circumstance where bodily harm is likely to occur without immediate action. Severe problems may be reported using a keyword such as "Mayday", indicating that a unit is actively involved in an emergency situation (e.g., a firefighter who is lost, trapped, severely injured, or out of air). The transmission of a "Mayday" should prompt the channel to be cleared of all non-urgent radio traffic so the Incident Commander can determine the location and status of the sender with the problem. Whatever the specific words selected, it is vital that all firefighters understand their relative urgency and the actions required for each. To prevent complacency, "Mayday" should be reserved for only the most exigent circumstances.

To improve fireground communication, fire departments should actively promote a culture in which it is acceptable to ask for help, clarify messages, and report problems. While following the chain of command is important operationally, it should be culturally acceptable to circumvent the chain for critical messages, when necessary.

Completing the Loop: Two-Way Communication

For communications to be effective they must follow the well-known "loop" model wherein the sender transmits a message and receives feedback from the receiver to ensure correct understanding of the message. Simply giving an order or information does not guarantee that it was received nor understood-the ultimate goal of communication. Effective, safe fireground communication is a two-way process. As demonstrated by several of the incidents mentioned here, it is vital that firefighters at every organizational level understand the importance of two-way communications for ensuring safety and sound decisionmaking on the fireground. Effective two-way communication has the added benefit of improving team cohesion and cooperation during routine daily activities. The purpose of this section is to provide recommendations for ensuring that two-way communication takes place.

There are a variety of related situations where firefighters can fail to "complete the loop". For example, a message may not be received at all by the intended recipient; the message may be received, but not understood; messages can be received and misunderstood; messages can be received, understood, and not acted upon; and messages can be received and deliberately, or selectively, ignored. For a variety of reasons, all of these situations may occur on the fireground or emergency scene. Just one such occurrence can prove deadly to firefighters and rescue personnel. Fire departments can take several steps to help prevent negative consequences from arising out of a failure to "complete the loop".

First, formal acknowledges should be required for every message. While people are unaccustomed to using formal acknowledgments during ordinary conversation, their use is vital when communicating critical messages where using non-verbal cues to avoid ambiguity is not possible. A recent *Wildland Firefighter Safety Awareness Study*[1] recommended the use of one of three levels of formal acknowledgment for every message. With little modification, these same levels may be recommended for use by structural firefighters.

Level 1 – Simple acknowledgment when receiving routine information. For example:

> Sender: "Engine 1 responding."
>
> Receiver: "OK, Engine 1."

Level 2 – Acknowledgment and feedback of key information. For example:

> Sender: "Engine 1, report to the 10th floor."
>
> Receiver: "Engine 1 copies, reporting to the 10th floor.:
>
> > Or
>
> Sender: "Engine 1, advance a 1-3/4" line into the fire; apartment."
>
> Receiver: "Engine 1 copies, advancing a 1-3/4" line into the fire"
>
> > Or
>
> Sender: "Engine 2, pickup my supply line at Maple and Sycamore."
>
> Receiver: "Engine 2 copies, pickup the line at Maple and Sycamore."

---

[1] Wildland Firefighter Safety Awareness Study-Phase II-Implementing Cultural Changes for Safety. (TriData Corporation: March 1998), p.xvii

Level 3 – Used to acknowledge more complex instructions. This level of acknowledgment may also provide an opportunity to conduct dialogue on interpreting the instructions, or to request clarification. The receiver may repeat the order on where to go and what to do, and clarify what is expected if unclear. For example:

Sender: "Engine 1, advance a 2-1/2" line into Exposure 4 and attack the cockloft fire. If unsuccessful within 5 minutes, pull out."

Receiver: "Engine 1 copies, advance a 2-1/2" into Exposure 4 and hit the cockloft. Pull out if we can't knock it in 5 minutes. Does this building have a truss roof?"

Or

Sender: "Engine 1, we now have reports of victims on the 7th floor. You want a second alarm?"

Receiver: "Engine 1 copies, victims on the 7th floor. Send a second alarm and 2 additional truck companies."

The sender (including dispatchers, command officers, and company-level officers) has the obligation to actively monitor the radio until acknowledgment is received and the sender believes the message has been fully understood by the recipient. If there is no response initially, the send can probe for understanding by asking questions like, "Did you copy?", or "Please acknowledge". If the sender believes the receiver is unclear about a vital piece of information, the sender must continue to probe until satisfied the message is completely and properly understood. When non-routine messages are broadcast to multiple receivers, acknowledgments should be obtained from all recipients in a prescribed order or through polling by the sender. The fact that one crew makes an acknowledgment does not ensure that all those affected have received the message.

Receivers have the responsibility to acknowledge messages and to request clarification if they did not completely understand. Failure to acknowledge from the receiver should be considered cause for concern and the sender should follow up to determine why confirmation was not communicated.

Dispatchers are a critical component of the communications loop. They must have an in-depth understanding of the fireground environment to ensure their ability to triage messages according to their importance and re-broadcast any vital messages to all of those enroute to, or at the scene. The importance of dispatchers is illustrated by the tragic experience in Memphis.

"The Fire Communications Bureau (FCB) personnel were not vigilant and/or alert to potential problems while monitoring radio transmissions as was desired. Private Bridges (one of the deceased firefighters) made four portable radio attempts to position 05 (the fireground frequency) of the communications frequencies. The tone of the caller (Private Bridges) should have alerted personnel in the FCB that a potential problem was present. A more experienced operator may not have accepted these radio transmissions as being routine. There was an additional telephone call to the FCB from a firefighter in an engine house not involved with 750 Adams. The firefighter had heard the four radio transmissions and contacted the FCB to inquire about them. Further attempts to determine the caller's (private Bridges) status should have been attempted after the telephone conversation."[2]

---

[2] Memphis Fire Services-Board of Inquiry, p. 31

Dispatchers, with their high-powered transmitters and remote location, are perhaps in the best position to monitor the welfare of firefighters operating on the scene of emergencies. Also, it is important that they do not override or "walk on" important messages sent from field units. Although a rare practice, some jurisdictions ensure maximum understanding by exclusively assigning fire department personnel as operational dispatchers for fire department communication systems. The Montgomery County (MD) Department of Fire-Rescue Services uses cross-trained fire department personnel to dispatch fire and emergency medical units. These individuals possess levels of training and experience comparable to their counterparts in the field, thus ensuring that communications personnel understand the complex inner workings of the fire or emergency scene.

SUMMARY OF RECOMMENDATIONS

- Ideally, all firefighters should be individually equipped with portable radios. At the very least, each two-person team entering a fire situation should have a portable radio.

- A better portable radio, suitable for and designed around use in the structural firefighting environment, is a priority need for the fire services.

- More training should be conducted to develop effective firefighter communication skills. These skills should receive a greater emphasis in training priorities.

- Policies and procedures should be developed that define: standard message format, important/urgent messages vs. routine messages, Mayday procedures, procedures for operations conducted on multiple channels, roles and responsibilities of those involved in the communications process at every level, and procedures for regular situation reporting.

- Radio discipline, while important must achieve a balance between limiting non-essential radio traffic and ensuring that potentially important information is regularly broadcast.

- The effectiveness of any incident management system depends on effective communication between firefighters and the Incident Commander. Chief officers may need aides to help monitor radio traffic during the incident.

- All firefighters should practice actively listening to radio traffic for information that may affect the performance of their assignments.

- Senior fire officers have a role to play in installing a department culture that encourages firefighters to request assistance and communicate operational problems. Rather than supporting a mentality that rewards excessive risk-taking, senior officers should emphasize that calling for help at the first sign of problems, is the expected action for safe emergency operations.

- Firefighters can reduce interference factors by turning down the volume on portable radios, shielding microphones, turning off sirens before transmitting when possible, maximizing face-to-face communications.

- Dispatchers should be continually involved in fireground communication by actively listening for transmissions that might go unnoticed, reporting changes to the normal response order, and conveying messages among responding units. Care should be taken, however, to minimize fireground radio traffic from being overridden by powerful transmitters at dispatch centers.

- More attention and research should be directed toward identifying barriers to effective communication and proactively preventing communication problems before an incident.